John Lewis-Stempel is a writer and farmer. His books include the *Sunday Times* bestsellers *The Running Hare* and *The Wood*. He is the only person to have won the Wainwright Prize for Nature Writing twice, with *Meadowland* and *Where Poppies Blow*. In 2016 he was Magazine Columnist of the Year for his column in *Country Life*. He farms cattle, sheep, pigs, and poultry. Traditionally.

The
Curious Life
of the
Cuckoo

John Lewis-Stempel

doubleday

TRANSWORLD PUBLISHERS
Penguin Random House, One Embassy Gardens,
8 Viaduct Gardens, London SW11 7BW
www.penguin.co.uk

Transworld is part of the Penguin Random House group of companies
whose addresses can be found at global.penguinrandomhouse.com

First published in Great Britain in 2025 by Doubleday
an imprint of Transworld Publishers

A CIP catalogue record for this book
is available from the British Library.

ISBN 9780857526410

Typeset in 11/14.5pt in Goudy Std by Jouve (UK), Milton Keynes.
Printed and bound in Great Britain by Clays Ltd, Elcograf S.p.A.

The authorized representative in the EEA is Penguin Random House Ireland,
Morrison Chambers, 32 Nassau Street, Dublin D02 YH68.

Penguin Random House is committed to a sustainable future
for our business, our readers and our planet. This book is made
from Forest Stewardship Council® certified paper.

Tempus adest veris, cuculus, modo rumpe soporem.

Ascribed to Bede (672–735)

CONTENTS

To the Cuckoo

O blithe New-comer! I have heard,
I hear thee and rejoice.
O Cuckoo! shall I call thee Bird,
Or but a wandering Voice?

While I am lying on the grass
Thy twofold shout I hear;
From hill to hill it seems to pass,
At once far off, and near.

Though babbling only to the Vale
Of sunshine and of flowers,
Thou bringest unto me a tale
Of visionary hours.

Thrice welcome, darling of the Spring!
Even yet thou art to me
No bird, but an invisible thing,
A voice, a mystery;

The same whom in my school-boy days
I listened to; that Cry
Which made me look a thousand ways
In bush, and tree, and sky.

To seek thee did I often rove
Through woods and on the green;
And thou wert still a hope, a love;
Still longed for, never seen.

And I can listen to thee yet;
Can lie upon the plain
And listen, till I do beget
That golden time again.

O blessèd Bird! the earth we pace
Again appears to be
An unsubstantial, faery place;
That is fit home for Thee!

William Wordsworth (1770–1850)

INTRODUCTION

'Cuckoo', Definition. *Oxford English Dictionary*
NOUN 1) Long-tailed, medium-sized bird, typically with a grey or brown back and barred or pale underparts. Many cuckoos lay their eggs in the nests of small songbirds. 2) informal: A mad person.
ADJECTIVE informal: Mad; crazy.
ORIGIN Middle English from Old French cucu, imitative of its call.
PRONUNCIATION CUCKOO/'KƱKU:/

I have seen a common cuckoo only twice in my life, by which I mean incontestable close-up views of the bird. Once, when I was swinging under my grandparents' damson tree, a cuckoo alighted on the rose trellis across the lawn, close enough for me to see the bright yellow rim of its eyes. Kindly, it also adopted the dipped querying pose of page 62 of *The Observer's Book of Birds*. This cuckoo, utterly disdainful of a seven-year-old, stayed put for a good minute, allowing me to imprint the scene on my mind.

My second definite sighting was more than forty years later, when I was farming pigs in a Herefordshire wood, not that far in fact, as the cuckoo flies, from my grandparents' old house. I was in a clearing: a bar-chested bird flew at speed over my head. And uttered *cuckoo* as it passed.

The common cuckoo, unlike Victorian children, is more often heard than seen. Despite that one specimen displaying itself for my childish delight in a country garden, the cuckoo is an elusive bird, which is one reason, despite a life largely outdoors, for the paucity of my confirmed spottings. Another problem with cuckoos is that, when on the wing, they do look like other birds, hawks and falcons especially. This resemblance to small raptors goes a little way to explaining why the ancients, having no understanding of bird migration, believed that the cuckoo of summer changed in winter into a kestrel/sparrowhawk/merlin. In Plutarch's *Life of Aratus*, the cuckoo asks the other birds why they flee from his sight, inasmuch as he is not ferocious; the birds answer that they fear in him the future sparrowhawk. Such fantastical ideas about migratory birds persisted all the way into the Darwinian nineteenth century. (Another venerable answer for the cuckoo's seasonal disappearance, one favoured by the English birdmen Francis Willughby and John Ray in *The Ornithology of Francis Willughby*, 1676, was that the cuckoo lay torpid in a hollow tree in cold times.)

Of course, the most obvious reason for my, and everyone else's, rare sightings of the common cuckoo is that it has become anything but common. According to the latest assessment from conservation groups, the number of cuckoos has declined by more than 65 per cent since the 1980s, a decline matched by other once-common farmland birds such as the lapwing and yellow wagtail. Today the cuckoo is 'Red Listed' by the IUCN, the International Union for the Conservation of Nature. The number of breeding pairs of cuckoos in Britain is currently around 15,000.

The cuckoos of my childhood may have been shy or misidentified, but their distinctive diphone call was the sound of spring. *Cuckoo-cuckoo* seemed to emanate from every spinney and hedge in Herefordshire in the 1970s, which was the last decade in which England was bird rich. We, like generations of kids before us, knew the country ditty:

> *Cuckoo, cuckoo . . .*
> *What do you do?*
> *'In April, I open my bill.*

In May, I sing night and day.
In June, I change my tune.
In July, far, far, I fly.
In August, away! I must.'

Arriving here from its long migration, the cuckoo was spring's sure and aural herald. People wrote to *The Times* to proclaim the news of the first cuckoo heard. Nowadays, across swathes of Britain any cuckoo, let alone the first, is an event worthy of record.

Of interest for its date of arrival, the cuckoo was, and is, an object of absolute fascination for its habits, particularly its objectionable depositing of its eggs in the nest of another bird . . . and the yet more objectionable, even downright homicidal habit of the young cuckoo of throwing out of the nest every occupant. Except itself. Such 'unnatural' behaviour, one of the strangest tales of the countryside, suggested that the cuckoo was either mad or bad. In more than one dialect and language 'gowk' means both 'cuckoo' and 'fool'. Gowk comes from the Old English *géac*, cognate with Old Norse *gaukr*; it is an old name for the cuckoo, still used locally in the North and Scotland. Down south the cuckoo was also 'gawk', leading to gawky (as in teenager) and even, by a stretch of the etymological imagination, geeky.

On April Fools' Day in the past, it was the done thing to send someone on a gowk's errand, bearing a message that read, 'This is the first of Aprile, Hunt the gowk another mile.' To tell someone 'You're cuckoo' is still to tell someone that they are crazy.

In English literature, Chaucer early identified the cuckoo as malefic, as opposed to loon. In his poem 'The Parlement of Foules' the bird is 'the cukkow ever unkynde', 'mordrer of the heysugge [hedge sparrow] on the braunche/That broghte thee forth, thou rewthe- less glotoun'; the Frenchman Clément Janequin, following suit, had this to say about cuckoos in his long poem 'Le Chant des Oyseaux', 1529:

> Begone, master cuckoo
> Leave our assembly
> Each of us gives you to the owl
> Because you are just a traitor.
> Cuckoo, cuckoo, cuckoo
> Betraying in each nest
> Laying your eggs without being summoned.

The cuckoo's 'betrayal' led the bird to grant its name to sexual unfaithfulness by human wives: 'cuckoldry'. A 'cuckoo in the nest', it follows, is an unwelcome intruder. Like a child fathered by another.

Singer and signaller of spring, murderous and mysterious in its mating habits, the cuckoo has attracted folklore, myth, music and literary allusion in abundance.

The bird's true biology is no less compelling. After all, how and why does it parasitically lay its eggs in the nests of others?

And how on Earth does the cuckoo fabricate eggs in the colours of its host?

Cloud Cuckoo Land

For once, an expression in the English language has a definite, ungainsayable origin. Cloud cuckoo land was coined by the fourth-century BC Greek playwright Aristophanes in *The Birds*. The play was first translated into English by the poet Henry F. Cary in 1824, which is the date 'cloud cuckoo land' entered the language.

CHORUS LEADER: *So what name shall we give our city?*

PISTHETAIROS: *Well, do you want to use that mighty name from Lacedaimon – shall we call it Sparta?*

EUELPIDES: *By Hercules, would I use that name Sparta for my city? No. I wouldn't even try esparto grass to make my bed, not if I could use cords of linen.*

PISTHETAIROS: *All right then, what name shall we provide?*

CHORUS LEADER: *Some name from around here – to do with clouds, with high places full of air, something really extra grand.*

PISTHETAIROS: *Well, then, how do you like this: Cloud-cuckoo-land?*

1848

The Cuckoo and the Starling

One day a Cuckoo, in his flight up and down,
Fell in with a Starling escaped from the town:
'Pray what is the talk?' he began with an air,
'Pray how do they speak of songs in the city?
Pray what do they think of the Nightingale there?'
'The whole of the town is in love with her ditty!'
'And pray what remark do they make on the Lark?'
'She is high in renown with half of the town'
'Indeed! Well, and as to the Blackbird?' 'He, too,
Is eulogized much, here and there by a few.'
'Well, now, I've to add that I'd feel very glad
If you'd tell me the various opinions that go forth
Respecting myself and my merits and so forth.'
'Why, that,' said the Starling, 'I hardly can do
For scarcely a soul ever talks about you.'
'Base ingrates! Well, then, as they grant me no praise,
I'll trumpet myself to the end of my days.'
So saying, away to the forest he flew,
And ever since then has been crying, Cuckoo!

Christian Gellert (1715–1769)

The Real Life of the Cuckoo

There is a story from Old Europe that, originally, the cuckoo was a baker's man, which is why he wears a dingy meal-sprinkled coat. And once, when God was blessing the dough in the oven, the cuckoo greedily pulled lumps out, crying every time *guk-guk*, or *look-look*. For this robbery of the poor, the Lord punished him by changing him into a bird of prey. Which he is not.

This tale, fantastic as it is, describes the physicalities of *Cuculus canorus* well. The ash-grey bird does seem to wear an apron with transverse dusty brown streaks. Or, at least, the fable describes the male bird and some females; a minority rufous-coloured female morph resembles the juvenile of the species. Whatever its morph, in flight, the common cuckoo, with its pointed wings, has the look of the smaller birds of prey, though the small head, thin bill and spatula-shaped tail

11

are identifiers. Also diagnostic is the famous springtime call of the male, usually rendered *cuckoo* in English but sometimes as *guk-guk*, hence the 'gawk'-, 'gog'-, 'gok'-type country names. Across Europe the bird is similarly named after the male's signature signal, so it is *cuco* in Spain, *cuculo* in Italy, *coucou* in France, *kukuck* in Germany, *kaki* in Finland, *kukushka* in Russia, *kakukk* in Hungary, and *koekkoek* in the Netherlands.

Whether rendered by humans as *cuckoo* or *kukuck*, the diphone notes, woodwindish and booming, of the male bird's call are made by the cuckoo opening the bill for the 'cuck' and closing it to form a sound chamber for the 'oo'. The resultant sound can seem very human, which is one reason among many for our immemorial fascination with the charismatic *Cuculus canorus*. The British ornithologist Reverend C. A. Johns, one of the last of the nation's parson naturalists, went so far as to claim that 'no bird in a state of nature utters a note approaching so closely the sound of the human voice as the Cuckoo'. Over the course of the breeding season, which is late spring and early summer, the male cuckoo's song changes, however, from *cuckoo* to a stuttery *cuck-cuck-oo* to silence. In the words of a Gloucestershire folk rhyme:

The cuckoo comes in April,
Sings a song in May;
Then in June another tune,
And then she flies away.

A sixteenth-century poem runs:

In Aprill, the koo coo can sing her song by rote:
In June, of tune she cannot sing a note;
At first, koo-coo, koo-coo sing still can she do;
At last, kooke, kooke, six kookes to one koo.

From the same era of English literature, Shakespeare has Henry IV tell his son that, as summer advances, the cuckoo's note no longer attracts notice as it did in April, having grown familiar: 'Was but as the cuckoo is in June, heard, not regarded'.

The call of the courting female cuckoo, meanwhile, is best described as a 'water-babbling' noise.

They were strange fictions, those old superstitions that the cuckoo metamorphosed into a wintery avian raptor, or hibernated inside a log (or even as a log), but then again, why shouldn't a bird so extraordinary as to raise its young in another bird's nest be capable of changing guise?

And isn't the real reason for the cuckoo's autumn and winter absence from Europe actually quite fabulous? The bird migrates thousands of miles to the jungles south of the Sahara. And then comes back again the following year.

In Britain, the cuckoo arrives in mid-April, traditionally on 14 April, St Tiburtius' Day. Until the 1940s *The Times* used to publish a letter every year from the first reader to report hearing a cuckoo. In the sprint to be the first recorder, over-enthusiasm occasionally occurred. From Mr Lydekker, F. R. S., 4 February 1913:

> *While gardening this afternoon I heard a faint note which led me to say to my under-gardener, who was working with me, 'Was that the cuckoo?' Almost immediately afterwards we both heard the full double note of a cuckoo, repeated either two or three times . . . There is not the slightest doubt that the song was that of a cuckoo.*

Seven days later, Mr Lydekker penned another letter to *The Thunderer*:

> *I regret to say that, in common with many other persons, I have been completely deceived in the matter of the supposed cuckoo of February 4. The note was uttered by a*

bricklayer's labourer at work in the neighbourhood of the spot whence the note appeared. I have interviewed the man, who tells me that he is able to draw cuckoos from considerable distances by the exactness of his imitation of their notes, which he produces without any kind of instrument.

Still, the earliest ever reliable record of a cuckoo in England was in February: one at Farnham, Surrey, on the twentieth of the month, 1953. Males arrive some two weeks before females in order to establish territory.

On its arrival, whenever it arrives, the cuckoo lives principally on beetles; when caterpillars, however, become abundant it prefers them, especially the hairy sorts. High on the cuckoo's menu of moths are:

Poplar hawkmoth
Elephant hawkmoth
Buff-tip
Antler moth
Yellow-tail
Lackey
Oak eggar
Drinker
Lappet
Small eggar
Figure of eight

White ermine
Buff ermine
Garden tiger
Cinnabar
March moth
Winter moth
Early thorn
Magpie
Mottled umber
Dotted border
Six-spot burnet

CUCKOO.

Hairy moth caterpillars are usually avoided by other insectivorous birds, due to their indigestibility. In contradistinction, the common cuckoo actually prefers its caterpillars to be bright, hairy and toxic. The bird is blessed with a specially developed digestive tract which can process the hairs and toxins of the caterpillars without harm; when the bird's stomach lining becomes a pincushion of spikes, it ejects the lining as a pellet and starts anew. *Cuculus canorus* is a woefully unacknowledged ecosystem manager: the cuckoo's foraging behaviour helps control the population of hairy moth caterpillars, some of which are noted plant destroyers, including of crops for the human table.

The common cuckoo is also an impeccable indicator of biodiversity; where there are cuckoos there are other birds (its hosts) and where there are cuckoos there are insects (its nutrition). Although the bird will hunt on the ground, its preferred modus operandi is 'perch and scan'. After filling its stomach, the male cuckoo turns its fancy to reproduction, booming out the distinctive *cuckoo* audio-lure to the female from a song post, perching majestically high in a tree, in a tail raised, wings dropped posture. Should a female be enticed into a visual assessment, the male performs a variety of stylized mating rituals and stratagems: the head is bobbed, the body bowed, wing held open

and drooped, the tail raised and fanned. While strut-
ting his stuff, the male may proffer nuptial gifts,
such as a piece of vegetation, or even, fantastically,
information – according to the journal *Animal Behaviour*, there is evidence that the male's dance moves are
more elaborate in areas where host birds are more
plentiful, which is essential knowledge for females.

CUCULUS canorus. *Gemeiner Kuckuk.*
1. W. m. Übergangskl.   2. zweijaehrig. W.

The cuckoo's mating ritual is oft repeated; both males and females are promiscuous, with multiple partners. A hen cuckoo may lay eggs of several sires. After courtship, nature takes its course and the female's egg-laying work begins.

In the jargon of ornithology, the common cuckoo is an 'interspecific brood parasite', meaning it lays its eggs in the nests of species other than its own, leaving the host to raise the cuckoo young, the original 'cuckoo in the nest'.

Across Europe, 145 bird species act as host to the common cuckoo. During the months of May and June, the female cuckoo lays her eggs, between twelve and twenty-five over the summer, all in different nests across the locality. If she fails to find a nest of her favourite host, as a last resort she might lay in the nest of another host species, but this egg is more likely to be rejected, as it will not be a decent facsimile of the host egg. If she fails to find a host species, she will, in desperation, lay her egg in any old bird's nest. In 1913, the Woolhope Naturalists' Field Club of Hereford-shire reported the presence of a cuckoo egg in a swallow's nest. The major cuckoo fosterers in Britain are meadow pipits (*Anthus pratensis*), reed warblers (*Acrocephalus scirpaceus*), pied wagtails (*Motacilla alba yarrellii*) and dunnocks (*Prunella modularis*). Other well-used hosts are: the robin, redstart, whitethroat, willow warbler, sedge warbler, skylark, wren, song thrush, chaffinch, goldfinch.

How does the hen cuckoo get away with the scandalous behaviour of laying eggs in another avian's nest? Cuckoos need an arsenal of trickery to get past host defences, for hosts are continually on the alert for cuckoo eggs and, if they detect one, they puncture it and eject it from the nest. If hosts espy a hen cuckoo they will mob it, attack it, beak and claws, and do real damage. Even kill it. The little birds know the nature of a brood parasite.

Until the early twentieth century, it was widely believed that the female cuckoo laid her egg on the ground, and deposited it in the nest of the host by carrying it in her bill. Thus the 1938 edition of the Reverend Johns' *British Birds in Their Haunts* declared:

In some of the instances, the position and structure of the nests were such that a bird of so large a size could not possibly have laid an egg in the usual way. Hence, and from other evidence, it is pretty clear that the egg is in all cases laid at a distance from the nest and carried by the bird in her bill to its destination. The bird can have no difficulty in accomplishing this seemingly hard task; for the gape of the Cuckoo is wide, and the egg

proportionately small, no larger in fact than the egg of a
skylark, a bird only a fourth of its size.

The truth had already been established by the British businessman and egg collector Edgar Percival Chance (1881–1955), who observed a cuckoo laying, confirming it did what most other birds do, but rather more quickly. Chance made a silent twelve-minute black-and-white film, *The Cuckoo's Secret*, on Pound Green Common in 1921, which showed to the world the female cuckoo's modus operandi: she lurks, statue-still, in a lookout tree, observing and listening; then, when a selected host has left to feed, she dives down and lays in its nest. She also devours one of its eggs – sometimes more – to give her own sufficient room (there is a limit to what the host bird can incubate under its body) and herself a free meal. Mission accomplished, she takes to the sky, never to lay eyes on her offspring: on average, the whole exercise takes a mere eight seconds. The fastest speed recorded by Chance on Pound Green Common for his main cuckoo subject, 'Cuckoo A', was four seconds; the longest sixteen seconds.

When the female cuckoo makes her swift exit, she often gives a 'chuckle', as if celebrating that her foul deed is done. This tittering is similar to the rapid

call notes of a sparrowhawk – which the bird has physically evolved to resemble – and it diverts the hosts'
attention away from noticing changes in their clutch
and towards their own safety instead. As a commando
raid, the cuckoo's laying combines both stealth and
deterrent.

But why does the cuckoo hen have to be so quick,
so secretive in the laying process? If the hosts are
alerted to the presence of a cuckoo near by they are
infinitely more suspicious of the eggs in their nest. If
they are unaware of the cuckoo, they will incubate the
interloper in blissful ignorance. Also, if the cuckoo is

noted by its victims during her preliminary loitering with intent, the ensuing furore is more likely to bring the nest to the attention of predators, putting the cuckoo's own egg in jeopardy.

A female cuckoo waiting for her opportunity to lay moves her egg down her oviduct so she can drop it virtually on demand. Equally, she can delay the laying of the egg for hours, although not for ever, as one Mrs Cecil Dawnay confirmed in a letter to *Country Life* in the summer of 1922:

On June 4, Whit-Sunday, while we were playing tennis on the lawn in front of our house, my nurse called from the balcony outside the nursery that a cuckoo had flown in and was sitting on the nursery floor. My sister went up and caught the cuckoo and brought it out on the lawn, where we admired it. It then gently fluttered on my small daughter's shoulder, and then, in the gravest manner possible, laid an egg, which fell unbroken to the ground. The cuckoo had been noticed previously on a tree near the nursery window, and as it had also been noticed that a pied wagtail had a nest a few feet away from the window, in a wisteria on the house, we imagined that the nest was one in which the cuckoo had intended to deposit one of its eggs.

A cuckoo's controlled laying – the holding of the egg for exact moment of deposit – can be extended to three hours, or thereabouts.

Cuckoos lay small eggs relative to their size. The egg of *Cuculus canorus* is 3.4g on average, whereas a thrush of the same body weight (100g) would be expected to lay a 10g egg. Since the cuckoo's egg is disproportionately small, carrying about the fully formed egg in her oviduct is unlikely to be a burden for the hen cuckoo as she waits for her moment to deposit her egg.

Clever birds, cuckoos. On Pound Green Common oologist Chance found that cuckoo females normally laid between 3 p.m. and 6 p.m. in the afternoon. Subsequently Nick Davies of the University of Cambridge, who has studied cuckoos on Wicken Fen in Cambridgeshire for quarter of a century, provided an explanation for the cuckoo's afternoon laying: the host is usually away feeding, having laid her own egg in the morning, so is less likely to detect and eject a counterfeit egg.

Over time, most host species have become adept at recognizing and ejecting parasite eggs. To avoid raising hosts' suspicions, cuckoo eggs have to closely mimic those of the host. It was once thought that a female cuckoo looked at the host's eggs, captured the image of them in its brain, then sent the image to the uterus

25

where it was reproduced on the surface of the egg she was about to lay. A biological colour scanner in other words. The fact of the matter is that, although female cuckoos all look alike, they belong to genetically distinct races, with each individual female cuckoo specializing in an exact host species – the host species she herself was born to. Thus reed warbler specialist cuckoos lay a greenish and spotted egg, just like those of reed warblers, while meadow pipit specialist cuckoos lay a brownish speckled egg, precisely like those of meadow pipits. In pied wagtail nests, the cuckoo egg is much paler, greyish white and finely speckled, like the wagtail's eggs.

Curiously, cuckoos that parasitize dunnock eggs have no need of a mimetic match. Cuckoo eggs in dunnock nests have a greyish-white background and reddish spots, unlike the plain-turquoise eggs of the dunnocks. Dunnocks just do not discriminate. Or are colour-blind, which is unlikely, because birds, courtesy of the extra colour cone in their eyes see in UV, as well as long (red), medium (green) and short (blue) wavelengths. Egg mimicry by cuckoos is superior in the most discriminating hosts. Dunnocks' pathetic lack of defences against cuckoos suggests that they are a species targeted relatively recently by *Cuculus canorus*.

There is evidence too that male cuckoos

contribute to egg mimicry: female cuckoos mate with a male of the same race. Probably, hens can tell cocks of their own genotype by voice. There may be more to that basic two-note *cuck-oo* call than mere humans can hear.

Evolution is not all in the cuckoo's favour. Hosts counter cuckoos by evolving telltale signatures on their eggs, which are difficult for the invading cuckoo to facsimile. The markings on the eggs of the great reed warbler, for example, differ drastically from female to female. The brambling, meanwhile, is among those species to develop egg patterns with a high degree of spatial complexity that are tough to replicate. Cuckoos and their hosts, in effect, are participants in an evolutionary, reciprocal 'arms race'. Or gene-for-gene coevolution, as it is properly, scientifically, termed.

Cuckoos lay on alternate days, since surveillance and commando egg-depositing raids are time and energy consuming. Their eggs hatch after an unusually short incubation period. On average, the cuckoo's egg needs just eleven days of host incubation, compared with twelve days for reed warbler eggs, twelve to thirteen days for dunnock eggs, and thirteen days for meadow pipit eggs. The early incubation comes about because – unlike most other birds – the cuckoo carries the egg, fully formed, in her oviduct for up to twenty-four hours, to

give her chick the head start, enabling it to hatch before the host eggs.

Cuckoo eggs are also thicker in shell than normal birds' eggs, to protect them from host attack, and to reduce damage during the speeded-up laying process. Cuckoos that parasitize birds with domed nests and narrow openings 'squirt' the egg in.

Very clever birds, cuckoos. Cuckoos will sometimes plunder the entire clutch of host species, causing the host avia to lay again. Consequently, laying by hosts is staggered throughout the cuckoo's bailiwick, and its own egg-laying opportunities maximized. Cuckoos are arch-manipulators.

While the cuckoo's egg mimicry is one marvel of

nature, another comes when the bird's chick is born. The sight of a petite pipit or warbler feeding an enormous cuckoo fledgling six times or more the size of the foster parent, has, in equal measures, astonished and appalled human observers for recorded time.

It was once believed, universally, that the eviction of the host's eggs and young must be a passive process, that they were simply squeezed out of the nest as the cuckoo chick enlarged. That the infant cuckoo casts out its host siblings *actively* was first noted by Edward Jenner, more famous as the pioneer of vaccination. (It would be more accurate to say that Jenner rediscovered this biological fact; Aristotle mentioned that the young cuckoo 'casts out of the nest those with whom it has so far lived' two thousand years before the Gloucestershire physician.) In 1788 Jenner published a paper on *Cuculus canorus* in the *Philosophical Transactions of the Royal Society of London*, the motto of which is *Nullius in verba*, 'Take nobody's word for it'. Jenner's paper was unpresumingly titled 'Observations on the Natural History of the Cuckoo'. This is Jenner's description of what occurred in a hedge sparrow's nest on 18 June 1787:

> . . . *to my astonishment [I] saw the young Cuckoo, though so newly hatched, in the act of turning out the very young Hedge-sparrow. The mode of accomplishing*

this was very curious. The little animal, with the assistance of its rump and wings, contrived to get the bird upon its back, and making a lodgement for the burden by elevating its elbows, clambered backwards with it up the side of the nest till it reached the top, where, resting for a moment, it threw off its load with a jerk, and quite disengaged it from the nest. It remained in this situation a short time, feeling about with the extremities of its wings, as if to be convinced whether the business was properly executed, and then dropped into the nest again. With these (the extremities of its wings) I have often seen it examine, as it were, an egg and nestling before it began its operations; and the nice sensibility which these parts appeared to possess seemed sufficiently to compensate the want of sight, which as yet it was destitute of.

I afterward put in an egg, and this, by a similar process, was conveyed to the edge of the nest and thrown out. These experiments I have since repeated several times in different nests, and have always found the young Cuckoo disposed to act in the same manner. In climbing up the nest, it sometimes drops its burden, and thus is foiled in its endeavours; but, after a little respite, the work is resumed, and goes on almost incessantly till it is effected.

[. . .] The singularity of its shape is well adapted to these purposes; for, different from other newly hatched birds, its back from the scapulae downwards is very

broad, with a considerable depression in the middle. This depression seems formed by nature for the design of giving a more secure lodgement to the egg of the hedge-sparrow, or its young one, when the young Cuckoo is employed in removing either of them from the nest.

Jenner's paper was met with dubiety. Eventually, though, he was avouched by the foremost ornithologists of the era, led by George Montagu (he who gave his name to Montagu's harrier).

The one cuckoo chick trick Jenner missed in his paper on cuckoo evictions is that they can occur while the hapless host hen is sitting, and requires shoving aside.

Such is the primeval, grotesque instinct of the cuckoo baby to eject both host eggs and host chicks, that it will not tolerate anything else in the nest. Even a single, placed pebble.

The heaving-out process is not risk free for the infant cuckoo; sometimes, in its elemental unconscious hegemony, it follows its foster siblings over the edge.

Why eject its siblings? The newly hatched cuckoo is necessarily smaller than the egg from which it progressed, but a full-grown cuckoo exceeds the dimensions of an entire brood of pipits; its growth, consequently, must be rapid and requires a gargantuan supply of food.

Aside from fratricide, the cuckoo chick, like its mother, has a ruse or two to ensure the promulgation of the species. In 1743 there appeared a German book entitled *Winged Theology: an attempt to inspire humankind to admiration, love and reverence for their creator by a close consideration of birds*. The author, J. H. Zorn, wrote: 'The young cuckoo cries as loud as a whole brood of host nestlings. The reason is to enhance the feeding by the foster parents.'

The percipient Herr Zorn was absolutely correct, as experiments nearly three hundred years later have proved. The infant cuckoo's loud and rapid begging calls sound like a whole brood of hungry host young, and this fools the foster parents into bringing as much food to the cuckoo chick as they would to an entire brood of their own. As the chick grows, so the intensity of its begging cry increases to make the hosts work harder, victims of what Darwin called 'mistaken instinct'. The host is hardwired to feed its hungry young, and so the aural deceit succeeds.

It is an oddity of evolution that the host birds reject cuckoo eggs but not cuckoo chicks.

Alone in the nest, the cuckoo has the sole attention of its foster parents, who will dart around the day long to feed it. The chick lives like a dauphin, an only child, a spoilt brat.

The infant cuckoo utilizes another wile in the nest, as well as brood-impersonation, but this time to ward off predators, be they other birds, reptiles or mammals. As Jenner observed in his 1788 paper:

Long before it leaves the nest, it frequently, when irritated, assumes the manner of a bird of prey, looks ferocious, throws itself back, and pecks at any thing presented to it with great vehemence, often at the same time making a chuckling noise like a young hawk. Sometimes, when disturbed in a smaller degree, it makes a kind of hissing noise, accompanied by a heaving motion of the whole body.

The orange gape makes the predator-impersonation especially effective, though the cuckoo in the nest's most effective anti-predator measure is the emittance of foul-smelling liquid brown faeces when touched.

Extremely clever birds, cuckoos.

Cuckoo chicks fledge from the host nest at around three weeks of age, by which time they have far outgrown their accommodation, though they continue to be fed, Billy Bunterish, on an adjacent branch by their poor foster parents for another two weeks. Then the juvenile cuckoos celebrate their independence by flying away, alone, to Africa in late July or August. They will never see their real parents; indeed, most adult cuckoos will already have left the UK by late June.

All of which leaves the age-old question about the cuckoo: is it a cruel and evil bird, Chaucer's 'mordrer of the heysugge'? Early nature writers, regarding the cuckoo in God's design, could make no sense of *Cuculus canorus* at all. The seventeenth-century naturalist John Ray wrote frustratedly:

> *The Cuckow her self builds no nest; but having found the nest of some little bird, she either devours or destroys the eggs she there finds, and in the room thereof lays one of her own, and so forsakes it. The silly bird returning, sits on this egg, hatches it, and with a great deal of care and toil*

*broods, feeds and cherishes the young Cuckow for her
own, until it be grown up and able to fly and shift for it
self. Which thing seems so strange, monstrous and absurd,
that for my part I cannot sufficiently wonder there should
be such an example in nature; nor could I have ever been
induced to believe such a thing had been done by nature's
instinct, had I not with my own eyes seen it.*

Edward Topsell concluded in *The Fowles of Heaven*,
1613, that the Creator made the cuckoo a parasite in
an act of beneficence; the cuckoo, because of her
'oune frigiditie, or coldness of nature' being unable to
hatch her own eggs, had them raised by others. 'Nature
being defective in one part is wont to supply by
another . . . the work of God is wonderful.'

Gilbert White (1720–93), the celebrated naturalist of Selborne, found no such goodness: the desertion
of the eggs by the cuckoo hen was a 'monstrous out-
rage on maternal affection'.

It was Charles Darwin who cut to the truth in
chapter eight of 1859's *On the Origin of Species*. He
identified the immense evolutionary benefit of being a
brood parasite, such behaviour by all means a com-
petitive advantage as opposed to deficiency. Being
freed of maternal duties means that cuckoos, whose
original occasional parasitism became full time, can

35

migrate earlier and with saved resources lay more eggs per season than their victims. Cuckoos are notably plentiful in their laying.

Darwin's observation prompts one question: if the cuckoo's con is so effective, why aren't more birds at it? There are 10,000 species of bird in the world, but only 102 are obligate brood parasites like the common cuckoo. A family DNA tree for cuckoos, constructed by Michael Sorenson and Robert Payne in 2005, demonstrates that even within the Cuculidae just half of the family are full-time parasites.

The drawback of brood parasitism is that, over evolutionary time, the victim erects ever more defences. To break these down requires effort. The cuckoo now has to work hard to be lazy.

Neither mad nor bad, the cuckoo is nature at work, in one of the most curious examples of Darwin's survival of the fittest.

With no more new host clutches to parasitize, their summer's job done, the adult cuckoos set off on their long journey south to Africa; adult cuckoos may sojourn in the UK for as little as four weeks. The young cuckoos follow them a few weeks later and all alone.

The cuckoo spends nine months of the year in tropical Africa. Where it has never been heard to sing.

Country Names for Cuckoo, *Cuculus canorus*

Gawk, Gowk (Dorset)
Geck, Gog, Gok (Cornwall)
Gowk (Yorkshire, Scotland)
Hobby (Norfolk)
Welsh Ambassador (West Country; since the
   bird seemed to arrive from the west, thus
   Thomas Middleton in his play *Your Five
   Gallants*, 1607, 'Your deuice here is a cuckow
   sitting on a tree, the Welsh lidger', with
   'lidger' a synonym for ambassador.)

The Cuckoo: The Only Bird that is Killed by Those of its Own Kind – A Bird that Lays Only One Egg

**Pliny, the Elder: From *Natural History*
(AD 77–79)**

The cuckoo seems to be but another form of the hawk, which at a certain season of the year changes its shape; it being the fact that during this period no other hawks are to be seen, except, perhaps, for a few days only; the cuckoo, too, itself is only seen for a short period in the summer, and does not make its appearance after. It is the only one among the hawks that has not hooked talons; neither is it like the rest of them in the head, or, indeed, in any other respect, except the colour only, while in the beak it bears a stronger resemblance to the pigeon. In addition to this, it is devoured by the hawk, if they chance at any time to meet; this being the only one among the whole race of birds that is preyed upon by those of its own kind. It changes its voice also with its appearance, comes out in the spring, and goes into retirement at the rising of the Dog-star. It always lays its eggs in the nest of another bird, and that of the ring-dove more especially – mostly a single egg, a thing that is the case with no other bird;

sometimes however, but very rarely, it is known to lay two. It is supposed, that the reason for its thus substituting its young ones, is the fact that it is aware how greatly it is hated by all the other birds; for even the very smallest of them will attack it. Hence it is, that it thinks its own race will stand no chance of being perpetuated unless it contrives to deceive them, and for this reason builds no nest of its own: and besides this, it is a very timid animal. In the meantime, the female bird, sitting on her nest, is rearing a supposititious and spurious progeny; while the young cuckoo, which is naturally craving and greedy, snatches away all the food from the other young ones, and by so doing grows plump and sleek, and quite gains the affections of his foster-mother; who takes a great pleasure in his fine appearance, and is quite surprised that she has become the mother of so handsome an offspring. In comparison with him, she discards her own young as so many strangers, until at last, when the young cuckoo is now able to take the wing, he finishes by devouring her. For sweetness of the flesh, there is not a bird in existence to be compared to the cuckoo at this season.

The Cuckoo

The cuckoo, like a hawk in flight,
With narrow pointed wings
Whews o'er our heads – soon out of sight
And as she flies she sings:
And darting down the hedgerow side
She scares the little bird
Who leaves the nest it cannot hide
While plaintive notes are heard.

I've watched it on an old oak tree
Sing half an hour away
Until its quick eye noticed me
And then it whewed away.
Its mouth when open shone as red
As hips upon the brier,
Like stock doves seemed its winged head
But striving to get higher

It heard me rustle and above leaves
Soon did its flight pursue,
Still waking summer's melodies
And singing as it flew.
So quick it flies from wood to wood

The Real Life of the Cuckoo

'Tis miles off 'ere you think it gone;
I've thought when I have listening stood
Full twenty sang – when only one.

When summer from the forest starts
Its melody with silence lies,
And, like a bird from foreign parts,
It cannot sing for all it tries.
'Cuck cuck' it cries and mocking boys
Crie 'Cuck' and then it stutters more
Till quick forgot its own sweet voice
It seems to know itself no more.

John Clare (1793–1864)

The Cuckoo's Calling

There truly is more to the male cuckoo's bisyllabic *cu-coo* than meets the human ear. This aural claim on territory and advert for a mate is individual to each bird; all *cu-coos* are not the same, although the first note uttered by a bird is always higher than the second, with the difference between them, musically, a perfect descending third. So particular and sensitive is the cuckoo's call that soil, habitat and the presence (or not) of cuckoos of the same specialism will affect its timbre. Neither is the iconic call always two syllables; on duetting with a female, a male may a opt for a three-syllable version (*cuckcuckoo*). Territorial males will call repeatedly during the day, and also into the night. The cuckoo behind our house will call as late as 11 p.m., and start again at 6 a.m.

The female bird is less mute than often supposed. The female's bubbling chuckle, imitative of the sparrowhawk and used to scare off potential hosts, can also be a love song to the male of the species. One study discovered that 67 per cent of male-female cuckoo duets were initiated by the hen bird's *kwik-kwik-kwik*.

Ode to the Cuckoo

Hail, beauteous stranger of the grove!
Thou messenger of Spring!
Now Heaven repairs thy rural seat,
And woods thy welcome sing.

What time the daisy decks the green,
Thy certain voice we hear:
Hast thou a star to guide thy path,
Or mark the rolling year?

Delightful visitant! with thee
I hail the time of flowers;
And hear the sound of music sweet
From birds among the bowers.

The school-boy, wandering through the wood,
To pull the primrose gay,
Starts the new voice of Spring to hear,
And imitates thy lay.

What time the pea puts on the bloom,
Thou fliest thy vocal vale –
An annual guest, in other lands,
Another Spring to hail.

Sweet bird! thy bower is ever green,
Thy sky is ever clear;
Thou hast no sorrow in thy song,
No Winter in thy year!

Alas! sweet bird! not so my fate;
Dark scowling skies I see
Fast gathering round, and fraught with woe
And wintry years to me.

O could I fly, I'd fly with thee!
We'd make, with joyful wing,
Our annual visit o'er the globe –
Companions of the Spring.

Michael Bruce (1746–67)

The Cuckoos of Europe

Common, or European, Cuckoo (*Cuculus canorus*)

Appearance: Grey head with a thin bright-yellow ring around their eye and a black beak. Wings are pointed, and often drooping when at repose. Tail is long with rounded end. Upper parts grey, with barring on chest. Female similar to male, though with some buff on breast observable close-up. Females occasionally take a rusty-brown form. Flight is fast, and low. Repeated cuckoo song of male, delivered from song post and from time to time on the wing, is the sure tell; female has a babbling call.

Size: Length: 32–34cm; Wingspan: 58cm; Weight: 110–130g

Migration: Breeds across whole of Europe, except for Iceland and parts of northern Norway. Outside Europe the common cuckoo's breeding range extends across two-fifths of the Earth's land surface, including much of Asia

eastwards to Japan, and south to China. Cuckoos that breed in the west of this transcontinental range winter in Africa, while those breeding in the east winter in southern Asia.

In the UK the cuckoo is protected under the Wildlife and Countryside Act, 1981. Classified in the UK as a 'Red List' species under the Birds of Conservation Concern review and as a Priority Species in the UK Biodiversity Action Plan.

Great Spotted Cuckoo (*Clamator glandarius*)

Appearance: Pale chest, and ash-grey cap and wings. Noisy, often giving krree krree krree calls, but no cuckoos.

Size: Length: 35–39cm, but looks much larger than the common cuckoo due to the breadth of its wings (58–60cm).

Migration: *Like* Cuculus canorus *it winters in Africa, arriving in Europe in February, departing July (adults). In Europe the bird breeds in the Iberian Peninsula, southern France, western Italy, Croatia, south-east Bulgaria, north-east Greece, Turkey and Cyprus; outside this continent the bird breeds in North Africa and the Middle East. A bird of semi-arid open country with trees, it feeds on insects, spiders, small reptiles and hairy caterpillars, which are distasteful to many birds. It parasitizes the magpie and carrion crow, laying its egg in the latter's nest even while the hen magpie is sitting. Unlike the common cuckoo the great spotted cuckoo's hatched chick does not evict the host's eggs, but young magpies/crows perish due to lack of food. Accidental visitor to the UK.*

Oriental Cuckoo (*Cuculus optatus*)

Appearance: Similar to the common cuckoo (and indeed the Himalayan cuckoo – cuculus saturatus), though slightly smaller in size. Also known as the 'Horsfield's cuckoo'.

Size: Length: 30–32cm; Wingspan: 51–57cm; Weight 73–156g. Adult females and juveniles occur in two morphs, grey and rufous. The call of the male oriental cuckoo, wholly different to that of the common cuckoo, consists of four or eight poo-poo syllables, not unlike the hoopoe's 'booming' on the ear.

Migration: The bird has a large breeding range in northern Eurasia, including western Russia and possibly Finland. Outside this region, the breeding grounds extend eastwards as far as Japan. Winters in southern Asia, and Australia.

Sonnet XIX

The merry Cuckow, messenger of Spring,
His trompet shrill hath thrice already sounded:
That warnes al lovers wayt upon their king,
Who now is comming forth with girland crouned.
With noyse whereof the quyre of Byrds resounded,
Their anthemes sweet, devized of loves prayse,
That all the woods theyr ecchoes back rebounded,
As if they knew the meaning of their layes.
But mongst them all, which did Love's honor rayse,
No word was heard of her that most it ought;
But she his precept proudly disobayes,
And doth his ydle message set at nought.
Therefore, O Love, unlesse she turne to thee,
Ere Cuckow end, let her a rebell be!

Edmund Spenser (1552/3–99)

Taking Wing: Cuckoo Navigation

The common cuckoo migrates mainly by night, and often at high altitude, between three and five kilometres above the ground, where it seeks strong following winds. The migration distance from the summer breeding ground to wintering quarters in the Congo rainforests can be more than 7,000 miles, including a crossing of the Sahara, the main ecological barrier, in one fifty- to sixty-hour continuous flight. All of which begs the question, how does a lone young common cuckoo, raised by foster parents of other species, find the Congo without the help of experienced conspecifics – birds of the same sort – namely their parents?

Put simply, cuckoos inherit a navigational map and an internal compass, but quite how these work remains suitably enigmatic. The birds likely orientate in response to the stars, geomagnetic signals (as do a whole range of taxa), but also olfactory clues – they may, in a sense, be led by the nose. Years of cuckoo research by the British Trust for Ornithology and other agencies show that cuckoo navigation is no mindless compass-clock mechanism – en route the bird interprets and adapts to smells, structures, winds and food availability. (Very, very clever birds,

cuckoos.) Equally, the route from breeding ground to wintering ground is not direct: cuckoos follow defined flight corridors, ancient, narrow and immutable.

On their return from Africa, cuckoos generally return to the area of their birth.

Cuckoo Music

It is a sort of paradox that the cuckoo, the most unmusical of birds, is so often portrayed in music. The song of the cuckoo is various, but it's the simple, distinctive courting call of *Cuculus canorus* that has captured composers' hearts over the years – just two notes in the interval of a major or minor third, though the call is remarkably resonant, as poet William Wordsworth noted in 'To the Cuckoo': 'Thy twofold shout I hear/ . . . /At once far off, and near.'

The 'twofold shout' is distinctly audible at distances greater than a mile. The cuckoo's real value to music, of course, is that it makes a simple cipher for spring. Perhaps the first example of the cuckoo in European music is the old English Rota, or Round, attributed to John of Fornsete, a monk of Reading, England, about 1226, which begins with the phrases:

Sing, cuccu, nu. Sing, cuccu.
Sing, cuccu. Sing, cuccu, nu.
Sumer is i-cumen in –

By the eighteenth century the cuckoo's call was foundation for entire so-called 'Cuckoo Concertos'. Amongst these was one by Antonio Vivaldi, the Italian violinist, the work being contained in his series of concertos illustrating the four seasons. Vivaldi's instrument of choice for 'cuckoo' was clarinet, as it would be for, among others, Mahler and Saint-Saëns. The other instrument commonly used to reproduce *cu-cu* is the organ.

As well as symbolizing spring, the cuckoo's call occasionally enters music as a sign of madness. The theme tune, famously, of the American comedians Laurel and Hardy was 'Dance of the Cuckoos', composed by Marvin Hatley, and later copyrighted as 'Ku Ku'. The original theme, recorded by two clarinets in 1930, was re-recorded with a full orchestra in 1935.

We hear spring and zaniness in the cuckoo's song but its fellow avia, especially host species, hear horror. For dunnock, reed warbler and the like, the cuckoo's call is the call of doom – the cuckoo's nature as a brood parasite is well understood by its fellow feathered creatures.

A Cuckoo Playlist

John of Fornsete (poss), W de Wycombe (poss), 'The
    Cuckoo Song', (also known as 'Sumer is i-cumen
    in', 'The Reading Rota'), *c.*1226

Thomas Weelkes, 'The Nightingale, the Organ of
    Delight', *c.*1600

Girolamo Frescobaldi, 'Capriccio sopra il Cucho',
    *c.*1624

Bernardo Pasquini, 'Toccata con lo Scherzo del
    Cucco', 1698

Antonio Vivaldi, Violin Concerto in A major, RV
    335, 'The Cuckoo', 1723

Louis-Claude Daquin, 'Le Coucou', from *Pièces de
    Clavecin*, 1735

George Frideric Handel, Organ Concerto in F major,
    'The Cuckoo and the Nightingale', 1738–9

John Frederick Lampe, 'Cuckoo Concerto', *c.*1740

Trad. English folk song, 'The Cuckoo', earliest known
    version 1780

Ludwig van Beethoven, Symphony No. 6 in F major,
    Op. 68, II, 'Scene by the Brook', 1808

A. Herzog, 'Cuckoo Polka', 1859

J. Philip Sousa, 'Cuckoo Galop', 1873

J. M. Ross, 'Cuckoo Rondo', 1874

Gaelic folk song, 'The Cuckoo in the Grove'
(Cuachag nan Craobh), first recorded 1879
Camille Saint-Saëns, 'The Cuckoo in the Depths of
the Woods', from *Carnival of the Animals*, 1886
Gustav Mahler, Symphony No. 1 in D Major, 1887–8
Frederick Delius, 'On Hearing the First Cuckoo in
Spring', 1912
Ottorino Respighi, 'Il cucù' from *Gli Uccelli*, 1928
Melvin Hatley, 'The Dance of the Cuckoo', 1930
Mike Oldfield, 'Cuckoo Song', 1977 (based on a
piece written by Michael Praetorius (1571–1621)
Babyshambles, 'Cuckoo', 2013

'The Cuckoo Song'

Sing, cuccu, nu. Sing, cuccu.
Sing, cuccu. Sing, cuccu, nu.

Sumer is i-cumen in –
Lhude sing, cuccu!
Groweth sed and bloweth med
And springth the wude nu.
Sing, cuccu!

Awe bleteth after lomb,
Lhouth after calve cu,
Bulluc sterteth, bucke verteth
Murie sing, cuccu!
Cuccu, cuccu.
Wel singes thu, cuccu.
Ne swik thu nauer nu!

The thirteenth-century round known as 'The Cuckoo Song', also known more formally as 'The Reading Rota' because the oldest manuscript copy was found at Reading Abbey, is as misunderstood as it is loved. The familiar, ever quoted line, 'Sumer is i-cumen in' means not 'coming in' but 'has come'.

Summer is *nu*. Now. As proved by the presence of *Cuculus canorus*.

The other common mistranslation is the line 'Bulluc sterteth, bucke verteth.' The bullock has indeed 'started up', risen from his grassy grazing, yet he is also announcing his reproductive desires and territorial designs. The Latin word *stertere* means 'to snore' and is etymologically a cousin to 'snorting'. The buck deer's 'ferteth' is usually translated as the sitting-room polite 'cavorting', but it is as it sounds. Farteth. The spring-summer diet of deer, rich and plentiful, makes the animal, a ruminant, spectacularly gaseous. The medieval English were an earthy race.

For cuckoo-ologists, the interest in the rota, a type of part-song, lies in the last line. The medieval English were also a superstitious race, believing that the cuckoo brought spring on its wings, and also that the bird's call represents the number of years you have left to live. By asking the cuckoo never to stop singing, the singer is hoping that, if the bird stays, the fair days – the days of warmth and crop growth – will stay too. Equally, the singer is likely hoping for a long life for himself, which gives a very human, very transcendent edge to the plea to the cuckoo to 'ne swik thu nauer nu'. To never stop now.

J Owen Sculp.

CUCKOW

The Cuckoo in Myth and Folklore: Messenger of Spring and Morality

Among the birds, none has gathered so much legend, myth, folklore as the cuckoo, save the owl. The two birds, of course, have similarities: the onomatopoeic name; a call that is almost human (or at least easy for humans to reproduce); the mystery of their lives. *Cuculus canorus*, though, has its personal distinctions and curiosities.

First and foremost the *cuckoo-cuckoo* call is associated with the turn of the seasons. The cuckoo's distinctive cry has been the harbinger of spring for centuries in Western Europe; the bird arrives here in early spring after its yearly migration from Africa; the adults leave Europe by July, so the birds' appearance is brief and thus noticeable.

The Ancient Greek poet Hesiod talked about spring in his poem 'Works and Days' (seventh century

BC) as the season 'when the cuckoo first cuckoos in the leaves of the oak/and brings joy to mortals on the boundless earth'.

This is the oldest reference to the cuckoo in European literature. By Hesiod's rules of husbandry, the cuckoo's song marks the growing rains of spring, enabling 'the late plower [to] vie with the early', meaning the farmer who ploughs three days after the cuckoo's arrival will rival the one who ploughed in the autumn.

Old German law actually designated the official start of spring by the set phrase *'wann der gauch guket'*.

In the Middle Ages people believed that cuckoos literally brought spring with them; it came borne physically on their wings, and with the mating call of the male bird all of spring would start apace and the land would grow the crops on which they relied.

By association there are a number of farming rhymes and lores which feature cuckoos, this one from Victorian Norfolk:

> *If the cuckoo lights on a bare bough*
> *Keep your hay and sell your cow*
> *But if he comes on the blooming May*
> *Keep your cow and sell your hay*

Germanic peasants believed that grapes would fail to ripen if the cuckoo sung after St John's Day, 24 June.

In the time when people lived by signs and portents of the weather, the Scots held that the frequent calling of the cuckoo was a portent of the fierce 'gawk storms'. 'In the month of Averil, / The Gawk comes over the hill, / In a shower of rain', runs the Highland refrain.

Outside Scotland, the cuckoo, a spring bird after all, was usually associated with the sun; in Hindu cosmology the bird is a preeminent solar symbol.

The significance of the coming of the cuckoo in Western Europe is indicated by the way in which its name is bestowed on other springtime sights. Lady's smock (*Cardamine pratensis*) is also cuckoo flower and so is ragged robin (*Lychnis flos-cuculi* – *flos-cuculi* being the flower of the cuckoo). Lords-and-ladies (*Arum maculatum*) is also cuckoo pint – cuckoo's pint or penis, because of the prominence of the spadix, an allusion to the cuckoo's supposed promiscuity. Then there is 'cuckoo spit', the English label for the lump of froth which encloses the young froghopper bug. In Switzerland the cuckoo spit is *guggerspeu*, and in Denmark *giogespyt*.

There are traditional dates for the arrival of the

cuckoo in Britain. In Sussex it sings on St Tiburtius'
Day (14 April), in Hampshire on the fifteenth; in
Worcestershire there is a saying that the cuckoo is
never heard before Tenbury Fair, 21 April, the same
day it is due in Yorkshire. 'Cuckoo Day' in Scotland is
the twenty-fourth of the month.

On the day the cuckoo was first heard, farm work-
ers were given the day off, the bird's arrival celebrated
by drinking 'cuckoo ale', beer or ale drunk from the
barrel in the wood or spinney where the cuckoo was
calling. Shropshire miners also claimed the privilege
of a 'Cuckoo Day' holiday.

Cuckoo Day was marked in some places by a 'Cuckoo Fair'. At Heathfield Cuckoo Fair (known locally as the Heffle Cuckoo Fair) in East Sussex the annual tradition was for an old woman to release a cuckoo from her basket, whereupon he 'flies up England carrying warmer days with him'. Rudyard Kipling memorialized the fair in 'Cuckoo Song':

> *Tell it to the locked-up trees,*
> *Cuckoo, bring your song here!*
> *Warrant, Act and Summons, please,*
> *For Spring to pass along here!*
> *Tell old Winter, if he doubt,*
> *Tell him squat and square – a!*
> *Old Woman!*
> *Old Woman!*
> *Old Woman's let the Cuckoo out*
> *At Heffle Cuckoo Fair – a!*
>
> *March has searched and April tried –*
> *'Tisn't long to May now.*
> *Not so far to Whitsuntide*
> *And Cuckoo's come to stay now!*
> *Hear the valiant fellow shout*
> *Down the orchard bare – a!*

Old Woman!
Old Woman!
Old Woman's let the Cuckoo out
At Heffle Cuckoo Fair – a!

When your heart is young and gay
And the season rules it –
Work your works and play your play
'Fore the Autumn cools it!
Kiss you turn and turn-about,
But, my lad, beware – a!
Old Woman!
Old Woman!
Old Woman's let the Cuckoo out
At Heffle Cuckoo Fair – a!

According to legend, the visitors to the fair at Marsden, West Yorkshire, sought to try to prolong the cuckoo's stay by building a wall around its nest. Similar tales of villagers penning the cuckoo in so as to continue summer can be found across Britain. The earliest is perhaps that concerning the people of Gotham, Nottinghamshire, collected in the sixteenth-century *Merrie Tales of the Mad Men of Gotham*, but certainly pre-dating that anthology:

On a time the men of Gottam would have pinned in the Cuckoo, whereby shee should sing all the yeere, and in the midst of the town they made a hedge round in compasse, and they had got a Cuckoo, and had put her into it, and said, Sing here all the yeere, and thou shalt lacke neither meat nor drinke. The Cuckoo as soone as she perceived her selfe incompassed within the hedge, flew away. A vengeance on her said they, 'We made not our hedge high enough.'

'Cuckoo pen' is a common field name, especially on hill tops, which may indicate where such hopeless attempts at imprisonment took place. Equally, 'gowk' standing stones and cuckoo place names may refer to places where other rituals concerning *Cuculus canorus* occurred. The bird was regarded in pagan, pre-Christian Europe as a male phallic symbol, and associated with numerous fertility rituals. A common riff in Scandinavian folk songs has the cuckoo carrying the wedding-nut to the nuptials.

There are many pieces of lore about what to do on first hearing the cuckoo. If this should be on St Tiburtius' Day, you should jingle the money in your pockets, spit and not look at the ground in order to secure financial content. If you are in bed, the cuckoo's call portends illness, unless you start running at once. In Germany the sound of a cuckoo during a meal signified a year of hunger to come. A Celtic superstition maintains that a long journey lies before you in the direction you are looking when you hear the cuckoo first in spring.

Should you be in Somerset or Northamptonshire you can ask for a reasonable wish. The belief that the number of cuckoo calls a maiden hears on Cuckoo Day indicates the number of years before marriage is common across Europe; a Swedish folk rhyme runs:

> *gokj gok, sitt pa quist (on bough),*
> *sag mig vist (tell me true),*
> *hur manga ar (how many years)*
> *jag o-gift gar (I shall un-given go)?*

Since no one witnessed the migratory disappearance of the cuckoo, it was supposed that the bird never died, and that it was always the same cuckoo that sung year after year in the same wood. In its immortality, it must have seen everything, and must have gathered great wisdom (again, like the owl). Such quasi-divinity led to the capability of prophecy. (According to Danish myths the cuckoo is so busy with its forecasts it has no time or energy to build a nest.) In Goethe's

'Oracle of Spring' the seer bird informs a pair of lovers of their approaching marriage and the number of their children. Another divination, first recorded in 1579, was that by looking inside one's shoe on Cuckoo Day one would find the colour of the hair of one's future spouse. If neither marriage nor children were relevant, then, across much of Europe, the number of cuckoo calls showed how many years left you before you died. Unlikely though it seems, there is actually a correlation between the number of repeats in the cuckoo's calls (syllables) and the longevity of the hearer. A recent Danish research project found that in cuckoo-abundant areas, such as north Denmark, farmers lived long and well. Biodiverse places are better for human health.

In the days when omens were observed, it was considered a matter of high import to hear the song of the nightingale before that of the cuckoo. Thus Chaucer says:

> *it was a commone tale*
> *That it was gode to here the Nightingale,*
> *Moche rathir [earlier] than the lewde Cuckowe singe*

Milton's sonnet 'To the Nightingale', from c.1629, refers to the same superstition, that hearing the cuckoo in spring before the nightingale foretells bad luck for a lover:

> *O Nightingale, that on yon bloomy spray*
> *Warblest at eve, when all the woods are still,*
> *Though with fresh hope the lover's heart dost fill,*
> *While the jolly hours lead on propitious May;*
> *Thy liquid notes that close the eye of day,*
> *First heard before the shallow cuckoo's bill,*
> *Porten success in love. O, if Jove's will [. . .]*

As harbinger of spring, the cuckoo was a 'timebird', one of the migratory avians that set the calendar and the clock for people before widespread horology. The bird's 'timeliness' is perhaps one reason it donated its name to 'cuckoo clock', developed by woodcarvers of south-west Germany in the 1700s. According to legend the first cuckoo clock was a providential accident: the Black Forest horologist Franz Ketterer, using a system of miniature bellows and whistles, sought the recreation of a cockerel's crow. But the two notes emitted were akin to those of *Cuculus canorus* and the rest is history and a German cottage industry. Unfortunately for this oft-quoted story, the first known description of a 'kuckuck' clock precedes Ketterer by more than half a century, dating back to 1629 when

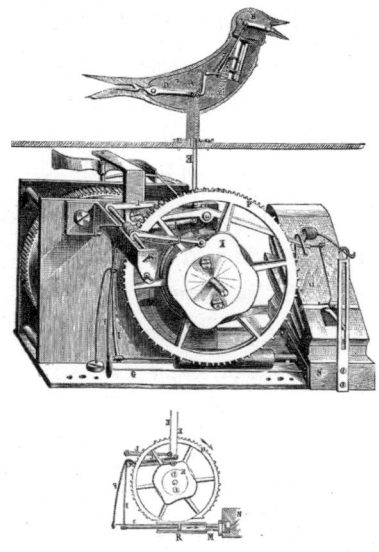

one was inventoried to Prince Elector August von Sachsen. Twenty years later, the Jesuit scholar Athanasius Kircher noted the workings of a mechanical cuckoo clock in an engraving in his compendium, *Musurgia Universalis*.

If the cuckoo is the messenger bird, it is also the bird of immorality, due to its brood parasitism, which contrasts diametrically with our notions of sound maternal instinct. Hence the 'cuckold', the human that suffers his marriage bed, his nest, 'adulterated' by another male. English poetry is rife with the association of cuckoos, betrayal and the planting of 'a cuckoo in the nest' – an illegitimate child. Shakespeare, famously, makes a word play between 'cuckoo' and 'cuckold' in *Love's Labour's Lost*:

> *When daisies pied and violets blue*
> *And lady-smocks all silver-white*
> *And cuckoo-buds of yellow hue*
> *Do paint the meadows with delight,*
> *The cuckoo then, on every tree,*
> *Mocks married men; for thus sings he:*
> *   'Cuckoo!*
> *Cuckoo, cuckoo!' O, word of fear,*
> *Unpleasing to a married ear!*

The cuckoo got sexed up early in human culture. In Greek mythology, the god Zeus transformed himself into a poor bedraggled cuckoo so that he could seduce the goddess Hera. In India, cuckoos became sacred to Kamadeva, the god of desire and longing. The god Indra in cuckoo-form (*kokilas*) seduced Rambhā – shades of Zeus and Hera – in the Sanskrit epic poem, *Rāmāyaṇam*.

So: the cuckoo is the betrayer bird, and cunning with it. Gustav Mahler's 1892 song cycle *Des Knaben Wunderhorn* (The Boy's Magic Horn), based on a traditional German folk verse, is an argument between a

nightingale and cuckoo over who is the finest singer. The cuckoo shows its guile by recruiting the donkey, with his 'two big ears, all the better to tell which is best'. The nightingale's song is too intricate for the donkey, who proclaims the elementary call of the cuckoo the winner.

There was no end to the sins of the cuckoo, the bad bird. One trans-European superstition asserted that the cuckoo could only stop singing once it had eaten its fill of wild cherries three times over. The Roman playwright Plautus (d. 184 BC) used the bird as an epithet for slothfulness – 'You cuckoo' – in *Pseudolus*, or *The Cheat*, because the cuckoo seems the laziest of birds, too idle to build its own nest. Shakespeare used it similarly in *Henry IV:* 'O' horseback, ye cuckoo'. 'Cuckoo' seems to have been in all ages an insult.

Sexy, sexually aberrant, promiscuous, shape-shifting, artful, lazy, gluttonous, the cuckoo joined cats and hares as the devil's work. Indeed, in medieval Europe, the cuckoo was Lucifer himself, hence the folk phrases 'cuckoo knows', 'cuckoo take him', 'cuckoo sent him'.

The cuckoo never fared well in Judaeo-Christianity, being designated forbidden food in the Old Testament, Leviticus 11:16, and the same prohibition is repeated in Deuteronomy 14:15, the Jewish people being ordered to hold the bird in abomination. The Ancient Greeks, it seems, had no such epicurean qualms, and their folk-belief ascribed portions of cuckoo as curative for stomach pains, and lumbago. The Romans, meanwhile, added cure for insomnia to the bird's list of positive attributes. According to Pliny, a cuckoo corpse bound with the skin of a hare – a lunar being – induces sleep. That is to say, the sun (symbolized by the cuckoo) hides itself, the moon (represented by the hare) appears, and the world (the human subject) falls asleep.

74

From 'The Cuckoo and the Nightingale', c.1380

Long attributed to Geoffrey Chaucer, more modern scholarship attributes the dream poem 'The Cuckoo and the Nightingale' to the Marcher soldier John Clanvowe. The poem, sometimes called 'The Book of Cupid, God of Love', is a debate between the birds on the subject of romance. There is no surprise in finding that the cuckoo is love's foe:

And as I lay, the Cuckoo, bird unholy,
Broke silence, or I heard him in my thought.

And that was right upon a tree fast by,
And who was then ill satisfied but I?
'Now, God,' quoth I, 'that died upon the rood,
From thee and thy base throat, keep all that's good,
Full little joy have I now of thy cry.'

And, as I with the Cuckoo thus 'gan chide,
In the next bush that was me fast beside,
I heard the lusty Nightingale so sing,
That her clear voice made a loud rioting,
Echoing through all the green wood wide.

'Ah! good sweet Nightingale! for my heart's cheer,
Hence hast thou stayed a little while too long;
For we have had the sorry Cuckoo here,
And she hath been before thee with her song;
Evil light on her! she hath done me wrong.'

But hear you now a wondrous thing, I pray;
As long as in that swooning-fit I lay,
Methought I wist right well what these birds meant,
And had good knowing both of their intent,
And of their speech, and all that they would say.

The Nightingale thus in my hearing spake: –
'Good Cuckoo, seek some other bush or brake,
And, prithee, let us that can sing dwell here;
For every wight eschews thy song to hear,
Such uncouth singing verily dost thou make.'

'What!' quoth she then, 'what is't that ails thee now?
It seems to me I sing as well as thou;
For mine's a song that is both true and plain,
Although I cannot quaver so in vain
As thou dost in thy throat, I wot not how.

'All men may understanding have of me,
But, Nightingale, so may they not of thee;

For thou hast many a foolish and quaint cry: –
Thou say'st O SEE, O SEE, then how may I
Have knowledge, I thee pray, what this may be?'

'Ah, fool!' quoth she, 'wist thou not what it is?
Oft as I say O SEE, O SEE, ywis,
Then mean I, that I should be wondrous fain
That shamefully they one and all were slain,
Whoever against Love mean aught amiss.

'And also would I that they all were dead,
Who do not think in love their life to lead;
For who is loth the God of Love to obey,
Is only fit to die, I dare well say,
And for that cause O SEE I cry; take heed!'

'Ay,' quoth the Cuckoo, 'that is a quaint law,
That all must love or die; but I withdraw,
And take my leave of all such company,
For mine intent it neither is to die,
Nor ever while I live Love's yoke to draw.

The Flight of the Cuckoo: The Resistible Decline of *Cuculus canorus*

C uckoo. The bird which has given its name to everything from adultery to the film *One Flew Over the Cuckoo's Nest* has, as noted, declined nationally by 65 per cent since the 1980s. And the good times were already over by then.

As befits an enigmatic bird, the cuckoo's population drop is a riddle itself. The fall of *Cuculus canorus* is undemocratically uneven. Indeed, contra the national picture, the Scottish population of the cuckoo has risen by 30 per cent; the Welsh population remains stable. At the micro level, within the squares of Ordnance Survey maps of England, it is clear that cuckoos are faring much better in heathland and hill than in farmed lowland.

The first culprit in the murder of the cuckoo, then? Intensive agriculture. With its hedge-ripping

and hedge-clipping, with its monotone multi-cut grassland, intensive agriculture has reduced the necessary habit for the cuckoo – and its preferred host species, the meadow pipit and the dunnock. No traditional old hay meadows, no rambly hedges with trees (which are needed, not least, by female cuckoos as lookout posts) means no meadow pipits, no dunnocks, which equals, by the law of nature, no cuckoos. The 'icides' of conventional farming, the pesticides, herbicides, the molluscicides and fungicides, have further reduced the food supply for the cuckoo (as well as its host birds). Specifically, the moth species whose caterpillars are a key food of adult cuckoos have plum-meted. Since 1968 the Rothamsted Insect Survey has monitored annual populations of large moths across England, Wales and Scotland, with a network of about one hundred light traps. In the thirty-five years to 2003, the annual total number of moths caught diminished by 31 per cent, with the greatest decrease in the southern half of the United Kingdom.

The moth effect on cuckoos is particularly pronounced because those hairy, toxic and aposematic moth caterpillars favoured by the bird – and avoided by most other avia – are highly sensitive to agricultural intensification. These moths, especially those in the families Lasiocampidae, Sphingidae, Notodontidae and Erebidae, have suffered declines of over 90 per cent in highly managed grassland and arable areas. In Devon, for instance, the cuckoo was widespread in the 1980s; it is now confined to the edges of Dartmoor and Exmoor. In Devon, and other similarly agro-industrial counties, the cuckoo has gone uphill to the high country. An adult cuckoo requires some 125 caterpillars daily down its gullet.

Culprit number two in the murder of the cuckoo is migration. As a brood parasite the cuckoo has a complex life cycle which, for UK birds, includes migrating more than 4,000 miles each spring from

81

Central and West Africa. Then returning there in summer. Only in the past decade has it been finally understood where cuckoos go in winter – to join the gorillas in the rainforests of the Congo. (It is a forgivable prejudice of the British to think of the cuckoo as 'our bird'; in fact, it spends 70 per cent of its life in the African jungle, not in Britain's mead, moor and spinney.) Up to that point in time, all the decades of ringing legs had produced just one clue: a cuckoo ringed as a fledgling in Eton on 23 June 1928 was felled by a local hunter's arrow in Cameroon, West Africa, on 30 January 1930. The bowman gifted the metal ring to his spouse as an ornament for her nose, the rhinal adornment then coming to the attention of the local pastor, who sensibly relayed the ring number to the British Museum.

Until very recently, most migratory birds have been too small to carry tracking devices. However, the British Trust for Ornithology's Platform Transmitter Terminals (PTTs), which weigh as little as 5g, now enable surveillance of the migrations and mortality of the cuckoo in close to real time.* Each individual is

---

* See page 89 for the British Trust for Ornithology's 'Sponsor a Cuckoo' scheme.

fitted with a PTT-100 backpack from Microwave Telemetry, connected to the bird by 2mm-diameter flat-braided nylon cord.

The migration routes of British cuckoos vary before the birds reach the great test of the Sahara, and account for something of the differing fortunes of our cuckoos on their return to Africa. British birds use two distinct routes during outbound migration: they either head south-west (the West route) via the Iberian Peninsula or south-east (the East route) via Italy, Greece and the Balkans. Most English cuckoos go south-west through Spain, but only half of the tracked birds survive the desert aerial trek. By contrast, most Scottish and Welsh cuckoos take a different summertide route, south-east through Italy; more than 95 per cent of these tracked cuckoos make it safely to winter quarters.

In other words, the birds taking the more deadly West route are from the already struggling English population.

What explains the excess mortality on the West route? Increasing droughts and wildfires in south-west Europe, together with habitat change, has made it hard for cuckoos using the route to maintain sufficient fat reserves for the long southing.

Culprit number three in the disappearance of the

cuckoo? Anthropogenic climate change. As global temperatures continue to rise, the cuckoo's host birds in the UK are breeding ever earlier. Therefore, northing cuckoos themselves need to arrive sooner for the best chance of breeding success. This is pushing cuckoos into precipitate departure from the important West African feeding stopover known as Intertropical Convergence Zone (ITCZ), where spring rains cause a bounty of insects. Effectively, cuckoos are trying the hazardous Sahara crossing under-fed and under-strength.

All in all, in England, at least, the cuckoo may not be i-cumen in for much longer.

Know, Nature, Like the Cuckoo, Laughs at Law

Know, Nature, like the cuckoo, laughs at law,
Placing her eggs in whatso nest she will;
And when, at callow-time, you think to find
The sparrow's stationary chirp, lo! bursts
Voyaging voice to glorify the Spring.

Alfred Austin (1835–1913)

Where to Hear (and See) Cuckoos in the UK

England: RSPB reserve at Lakenheath Fen (Suffolk), the New Forest, Dartmoor, Cannock Chase (Staffordshire), the North York Moors, Bough Beech (Kent, an important staging post for migrating birds), RSPB reserve at Cliffe Pools (Rochester, Kent), Trinity Broads (Norfolk), Clouts Wood (Wiltshire), Irthlingborough Lakes and Meadows (Northamptonshire), Glapthorn Cow Pastures (Northamptonshire), Highgate Common Nature Reserve (Wombourne, West Midlands), RSPB Arne (Dorset)

Wales: Tylcau Hill (Radnorshire), Brecon Beacons, Tregaron (Ceredigion), Mawddach Valley (Gwynedd)

Scotland: Everywhere outside the major conurbations and lowland agriculture of the east coast, but especially western Highlands, the Isle of Skye and the Hebrides

Northern Ireland: Ballynahone Bog (County Londonderry), Murlough National Nature Reserve (County Down)

Useful Addresses

British Trust for Ornithology (BTO)
The Nunnery, Thetford, Norfolk IP24 2PU
Tel: +44 (0)1842 750050
https://www.bto.org/
The BTO runs a 'Sponsor a Cuckoo' project, which
    enables you to track your sponsored bird in return for
    a donation. Funds from the project enable the con-
    tinuation of research into the bird and how it can
    be saved. https://www.bto.org/our-science/projects/
    cuckoo-tracking-project/get-involved/sponsor-
    cuckoo

The Wildlife Trusts
The Kiln, Mather Road, Newark, Nottinghamshire
    NG24 1WT
Tel: +44 (0)1636 677711
https://www.wildlifetrusts.org/

BIBLIOGRAPHY

Edward A. Armstrong, *The Folklore of Birds* (Collins, 1958)

Edgar Chance, *The Cuckoo's Secret* (Sidgwick & Jackson, 1922) [The film of the same name was rereleased by British Instructional Films on DVD in 2010.]

—*The Truth About the Cuckoo* (Country Life, 1940)

Simone Ciaralli et al, 'Ritual Displays by a Parasitic Cuckoo: Nuptial Gifts or Territorial Warnings?' in *Animal Behaviour*, 207 (2024), https://doi.org/10.1016/j.anbehav.2023.11.003

Nick Davies, *Cuckoos, Cowbirds and Other Cheats* (T & AD Poyser, 2011)

—*Cuckoo: Cheating by Nature* (Bloomsbury, 2015)

Chloe Denerley et al, 'Breeding Ground Correlates of the Distribution and Decline of the Common Cuckoo *Cuculus canorus* at Two Spatial Scales' in *Ibis*, 161:2 (April 2019), https://doi.org/10.1111/ibi.12612

J. E. Field, *The Myth of the Pent Cuckoo: A Study in Folklore* (Elliot Stock, 1913)

Angelo de Gubernatis, *Zoological Mythology* (1872)

Chris M. Hewson et al, 'Population Decline is Linked to Migration Route in the Common Cuckoo' in *Nature Communications*, 7:12296 (2016)

Reverend C. A. Johns, *British Birds in Their Haunts* (1862)

Orlando A. Mansfield, 'The Cuckoo and Nightingale in Music' in *The Musical Quarterly*, 7:2 (April 1921)

G. Montagu, *Ornithological Dictionary of British Birds* (1831)

F. Morelli et al, 'The Common Cuckoo is an Effective Indicator of High Bird Species Richness in Asia and Europe' in *Scientific Reports*, 7:4376 (2017), https://doi.org/10.1038/s41598-017-04794-3

Jeremy Mynott, *Birdscapes: Birds in Our Imagination and Experience* (Princeton University Press, 2009)

C. Moskát, M. E. Hauber, 'Male Common Cuckoos Use a Three-note Variant of their "Cu-coo" Call for Duetting with Conspecific Females' in *Behavioural Processes*, 191 (October 2021), https://doi.org/10.1016/j.beproc.2021.104472

David Nance, 'Gouk Stones and Other Cuckoo Place-names: Prehistoric Cult Sites' in *Scottish Geographical*

Journal, 135 (July 2019), https://doi.org/10.1080/14
702541.2019.1635265

Robert B. Payne, *The Cuckoos* (Oxford University Press, 2005)

John Pollard, *Birds in Greek Life and Myth* (Thames & Hudson, 1977)

Anders Pape Møller et al, 'Cuckoo Folklore and Human Well-being: Cuckoo Calls Predict How Long Farmers Live' in *Ecological Indicators*, 72 (January 2017), https://doi.org/10.1016/j.ecolind. 2016.09.006

C. H. Poole (ed.), *A Treasury of Bird Poems* (1911)

Kasper Thorup et al, 'Flying on their Own Wings: Young and Adult Cuckoos Respond Similarly to Long-distance Displacement During Migration' in *Scientific Reports*, 10:7698 (2020), https://doi. org/10.1038/s41598-020-64230-x

Edward Topsell, *The Fowles of Heaven* (1613)

W. Verboom, 'Bird Vocalizations: the Cuckoo (*Cuculus canorus*)' (2009)

PICTURE CREDITS

The Cuckoo

The cuckoo's a bonny bird
He sings as he flies;
he brings us good tidings:
he tells us no lies.

He drinks the cold water
To keep his voice clear;
And he'll come again
In the spring of the year.

Traditional Scottish rhyme

If you enjoyed *The Curious Life of the Cuckoo*, you will love these previous books from John Lewis-Stempel:

THE SECRET LIFE OF THE OWL

There is something about owls. They are creatures of the night, and thus of magic. But – with the sapient flatness of their faces, their big, round eyes, their paternal expressions – they are also reassuringly familiar. We see them as wise, like Athena's owl, and loyal, like Hedwig. Here, John Lewis-Stempel explores the legends and history of the owl. And in vivid, lyrical prose, he celebrates all the realities of this magnificent creature, whose natural powers are as fantastic as any myth.

THE GLORIOUS LIFE OF THE OAK

The oak is our most beloved and most familiar tree. For centuries, oak touched every part of a Briton's life – from cradle to coffin. It was oak that made the 'wooden walls' of Nelson's navy, and the navy that allowed Britain to rule the world. John Lewis-Stempel explores our long relationship with this iconic tree and retells oak stories from folklore, myth and legend – oaks bearing the souls of the dead, the Green Man, and fertility rites on Oak Apple Day. Of all the trees, it is the oak that speaks most clearly to us.

THE PRIVATE LIFE OF THE HARE

The hare is a rare sight for most people. We know them only from legends and stories. They are shape-shifters, witches' familiars and symbols of fertility. They are arrogant, as in Aesop's 'The Hare and the Tortoise', and absurd, as in Lewis Carroll's Mad March Hare. In the absence of observed facts, speculation and fantasy have flourished. But real hares? What are they like? In elegant prose John Lewis-Stempel celebrates how, in an age when television cameras have revealed so much in our landscape, the hare remains as elusive and magical as ever.

THE WILD LIFE OF THE FOX

The fox is our apex predator, our most beautiful and clever killer. We have witnessed its wild touch, watched it slink by bins at night and been chilled by its high-pitched scream. And yet we long to stroke the tumbling cubs outside their tunnel homes and watch the vixen stalk the cornfield. Foxes captivate us like no other species. Exploring a long and sometimes complicated relationship, *The Wild Life of the Fox* captures our love – and sometimes loathing – of this magnificent creature in vivid detail and lyrical prose.

THE SOARING LIFE OF THE LARK

Skylarks are the heralds of our countryside. Their music is the quintessential sound of spring. The spirit of English pastoralism, they inspire poets, composers and farmers alike. In the trenches of World War One they were a reminder of the chattering meadows of home. We watch as they climb the sky, delight in their joyful singing, and yet we harm them, too. *The Soaring Life of the Lark* explores the music and poetry, the breath-taking heights and the struggle to survive of one of Britain's most iconic songbirds.

NIGHTWALKING

As the human world settles down each evening, nocturnal animals prepare to take back the countryside. Taking readers on four walks through the four seasons, John Lewis-Stempel reveals a world bursting with life and normally hidden from view. Out beyond the cities, it is still possible to see the night sky full of stars, or witness a moonbow, an arch of white light in the heavens. It is time for us to leave our lairs and go tramping. To join our fellow creatures of the night.